Beach Weekend Homes

Beach Weekend Homes

Author
Cristina Montes

Editor
Paco Asensio

Photos:
© Alan Weintraub: **Part of the landscape**
© Adriano Brusaferri: **A passion for what's natural**
© Teisseire Laurent (ACI Roca-Sastre): **A terrace overlooking the sea**
© Morel Marie Pierre (ACI Roca-Sastre): **A token of simplicity**
© Ricardo Labougle: **The beauty of the imagination, Facing the ocean,**
   With the ocean at one's feet, Maritime exoticism, Natural fusion,
   A sensory exercise, Between sea and sky, Alone beside the sea,
   The southern lighthouse, Visual dialogue, Bathed in light, Mediterranean
   inspiration, A return to tradition
© Pere Planells: **Domesticated shapes, Architectural synthesis,**
   White geometry

Translation
Bill Bain

Editing and proof reading
Wendy Griswold and Julie King

Art director
Mireia Casanovas Soley

Graphic design and layout
Emma Termes Parera

2001 © Loft Publications S.L. and HBI,
an imprint of HarperCollins Publishers

First published in 2001 by LOFT and HBI,
an imprint of HarperCollins Publishers
10 East 53rd St. New York, NY 10022-5299

Distributed in the U.S. and Canada by Watson-Guptill Publications
770 Broadway New York, NY 10003-9595
Telephone: (800) 451-1741 or (732) 363-4511 in NJ, AK, HI Fax: (732) 363-0338

Distributed throughout the rest of the world by
HarperCollins International
10 East 53rd St. New York, NY 10022-5299
Fax: (212) 207-7654

Hardcover ISBN: 0-06-620937-4
Paperback ISBN: 0-8230-0479-1
D.L.: B-21.848-2001

Printed in Spain

If you want to make any suggestions for work to be included in our forthcoming books, please e-mail us at
loft@loftpublications.com

Introduction

Nowadays, having a relaxing retreat where we can unwind, away from the hustle and bustle and the everyday routine, where we can restore our energies before going back to the grind, has become almost a must. This space, where we can relax with our families or intimate friends, can take countless forms, but is always centered around a place to spend those precious, hard-earned, weekend leisure hours.

The appearance of this second home is not what counts. Its size and location are not crucial factors. What matters is that it is a great place to live, somewhere we yearn to be when we are not there, an attractive, practical, private domain, full of vitality, even though we will never spend nearly as much time there as in our principal home.

This book takes the reader on an enchanting tour through a series of weekend retreats, all located near the sea. However, it is not intended as merely a stunning collection of photos. The objective is to offer practical, effective solutions for anybody with a second residence who wants to remodel it or simply make some changes. In addition, the book is full of ideas for anybody who is thinking of buying a weekend home in which to spend precious leisure time.

Every house featured in this book, whether on the shores of the Atlantic, the Mediterranean, or the Pacific, has its own history. It is these memories, combined with the homes' basic features and amenities, which demonstrate that the real pleasures of life are within our reach.

The projects included in this book span the globe, in many different climates and topography. Compact and exquisitely austere houses, suitable for simple, natural lifestyles, take their places beside houses literally bathed by the waters of the Atlantic and architectural gems whitewashed by the Mediterranean sun. The possibilities are endless, just like the residents' needs and tastes.

Before they bought or constructed their homes, the owners dreamed of being better able to enjoy their leisure time by finding new ways to live in close contact with the sea stretching out to the horizon.

Thus inspired, they sought a home where they could escape from the rush and noise of the world for a few days, slowing everything down and leaving the pressures behind them. In one's own world, life's little pleasures can be savored: having dinner as the sun sets, listening to the waves lapping on the shore, or just wondering, throughout the day, at the ever-changing nature of the sea.

Cristina Montes

The beauty of the imagination

Straight lines distinguish the exterior areas
and the spacious, light-filled, airy rooms of this
house, open to the sea.

Punta del Este, Uruguay

Perched on the Atlantic, along Uruguay's white-sand beaches, this house by architect Martín Gómez commands a view of the entire bay. Three porches–two at ground level and one on the upper floor, with a belvedere containing five large windows–soften the severe, angular architecture, creating a striking effect. Spacious rooms are bathed in natural light in the spirit of the urban loft. The key to the spectacular treatment of the living room lies in the integration of different spaces through the use of massive steel beams that also support the old adobe walls

The wooden deck, an area for rest and relaxation which employs a wide variety of furniture as sun loungers, cleverly adjusts to the slope of the land.

remaining from the original building, which the new structure has replaced.

The colors and fabrics unify the different styles in the house, while ingenious decorating solutions have been employed in the flooring and the walls, especially in the baths and kitchen.

The flooring throughout is beige-tinted polished cement broken by thin glass strips, giving the appearance of a surface composed of large tiles. Pale gray granite from a local quarry is used in the doors and windows, affording an original yet subtle atmospheric touch.

A bare minimum of furniture and a careful combination of styles, from the purely functional to older, restored pieces, blend in a space dominated by a range of neutral whites.

Serenity, assured aesthetics, and eclecticism combine very successfully in this building that exudes freedom.

Recycled pieces like this low, massive wooden table go side-by-side with much lighter furniture. Outdoors, the mixed styles result in charming visual interplays.

This staircase, white like the rest of the room, connects the two floors of the house. The space below it houses a uniquely designed, compact bathroom.

The master bedroom, upstairs, leads out onto the terrace. Restored pieces and furnishings of various styles establish a welcoming ambience which, as in the other rooms, maintains the harmonious white.

Facing the ocean

Striking architectural touches set off this house, isolated from the hubbub, alone with the sound of the waves.

Punta del Este, Uruguay

In a desert-like spot a few kilometers from Punta del Este, Uruguay, with the Atlantic at his feet, architect Pachi Firpo designed a dwelling that would merge with the surrounding landscape and be easily isolated from the rest of the world.

Two geometric structures make up the exterior of the complex. The first is rectangular and contains most of the rooms, while the other one is almost square and serves as the guest quarters. The two structures are joined together by a large porch terrace set off by logs. Tying everything together this way also breaks up the building's straight lines and solemnity, while making it dynamic and novel. The rectilinearity was also relieved by breaking the front of the main structure with large openings offering truly incredible panoramas. These "glass walls" provide a constant link between inside and outside, and make views of the surroundings possible from every conceivable corner.

The contrast between the dry landscape and the ocean can be enjoyed from the porch/terrace, an open space with warm furnishings in which wood and natural colors predominate.

The load-bearing elements, including the floors, utilize curupay, an exceptionally hardy wood from Paraguay. The cement walls have been painted an ochre reminiscent of the beaches, to completely integrate the architecture into its natural setting.

The interior decor is the product of a collaboration between the Argentinian decorator Laura Ocampo and the architect. They selected a distribution pattern typical of urban lofts. The simplicity and brightness that predominate outside are retained inside; the furniture is in keeping with a subtly contemporary classical style characterized by a strict purity of line, along with containment and harmony. Often, the delicacy and elegance contrast with the arid environment in which the house is located.

The furniture in the master bedroom is elegant and classical. At one of the large windows is a reading area with two armchairs, a matching bench, and two floor lamps. The natural light can be diffused, when necessary, by the reed window blinds.

The windows are distinctively distributed in an unconventional, symmetrical pattern. In the bathroom, a large picture window–free of blinds, as in the majority of the rooms–frames the landscape. A round–footed tub, wooden towel rack, and low Oriental table are the only decorative elements.

Decorative details: fabrics and flowers provide the chromatic note in elegant, refined interiors achieved with a neutral white. The perfect combination of styles, while dominated by a certain aesthetic classicism, lend character to the setting.

With the ocean at one's feet

This blue house in a glorious natural setting invites you to relax.

La Jolla, California, USA

Open space designed for pleasure was the goal for this building beside the ocean.

Several terraces partly covered with natural plant materials set off a structure clad in indigo, a color which is more typical of another kind of building, but which in this case affords an attractive, unique, visual connection with the marine environment.

From the top of the facade a back porch emerges like an ornamental crown of natural gray wood. This creates a light geometric detail that breaks up the formal linearity of the facade. Everything lends itself to visual play. The contrasts–color, form, and texture–are constant and successful, both inside and out.

A cornucopia of styles blend harmoniously throughout the interior of the house. Colonial furniture is combined with other, more rustic pieces; windows with elegant drapes join others left bare or covered by simple cotton blinds. These rooms employ extremely effective decorative solutions. For example, one of the polished wooden beams serves as a sturdy curtain rod. A stunning piece of natural linen covers the doorway between the living room and the main entrance, which leads directly to the flagstoned garden. There, one of the most original color contrasts can be found: two large, varnished wooden doors against a blue background, meeting the green lawn, a striking effect achieved through an eclectic combination of materials and colors.

A porch is used as an open-air dining room with magnificent ocean views.

The rooms succeed each other as subtly differentiated settings. Each space radiates its own spirit while maintaining a continuity throughout.

There is nothing garish; there are no pyrotechnics. The structure is defined by its attractive architectural framework. The decor, making obvious use of the warm setting, is based on a sophisticated rustic approach.

Combinations of materials, textures, and colors are skillfully employed to give character and personality to this house, noteworthy for its restrained architecture.

The flagstones mark the boundary of the living area, beyond which are vegetation and ocean. The furniture on this improvised terrace consists of a low table and two benches made of rough-hewn wood, treated to withstand the weather.

Inside the house, the real star is eclecticism. Combinations which in other circumstances would be unthinkable are highly effective here. The result is an elegant space, filled with character.

The simplicity and restraint of the various materials and objects provide attractive contrasts. These decorative yet natural everyday objects break up the living room's solemnity.

Maritime exoticism

Open to the outdoors and perfectly integrated into its welcoming setting, this restrained, geometrical building is a peaceful place for rest and relaxation.

Punta del Este, Uruguay

The first sign of the serene character of this house on the South American coast is the rustic porch running along the back.

Crowned by a leafy climbing plant that softens the architectural mass and contrasts strikingly with the soft pale vanilla of the exterior walls, the peaceful porch has everything needed for living outdoors during the hottest times of the year. A spacious wooden table and cane chairs the same color as the building's exterior suggest never-ending nighttime conversations. And during the day, the novel divan-like deck chairs, covered in a tough, blue-and-white striped cotton print that imparts a festive maritime touch, await anyone wishing to bathe in the sun.

Inside, exotic cane blinds filter the blazing light and create a calm, carefully-tended ambience. The informal dining area, in front of the large porch window, is marked off by a pastel blue fiber mat from one of the open-air markets in the area. In the same room, with no visual barriers, two large sofas are arranged around a wooden table in the colonial style to suggest a salon.

The porch that surrounds the building provides truly spectacular views of the ocean. The simple furniture makes it an ideal spot from which to enjoy the refreshing sea air.

The dominant color in the master bedroom is light blue. The bed is opposite a fireplace with subtle lines. A large carpet in warm colors completes the ensemble, inviting the viewer to sit down and read, relax in front of the fire, or just listen to the strong coastal winds.

All in all, it is a creation with pure, simple lines, in which personal style prevails.

Lights set between the porch's equidistant, load-bearing columns illuminate the area at night.

All the furniture in the house is simple and highly functional. The brick fireplace opposite the bed in the master bedroom makes it cozy during cold winter nights. Beside the hearth, two stripped-wood chairs create a practical reading nook.

The cane-roofed porch is covered by a climbing plant to enhance heat insulation. The green leaves contrast with the blue of the divans' upholstery–the only note of color in the otherwise neutral grouping.

Natural fusion

This simple house meets the ocean calmly.
Here, the passage of time is measured by the sun

Zihuatanejo, Mexico

Between the Atlantic Ocean and a leafy eucalyptus forest in Mexico is a house, built entirely of wood, called La Ola (The Wave), an allusion to the soft curves of its architecture. Architect and designer Mario Connio, in a brilliant interpretation of the owners' desires, carried out a detailed study to determine the orientation which would make the most of the setting's natural light.

Located at the very edge of the ocean, the house was conceived basically as a family refuge, to be used at any time of the year for painting, reading, writing, relaxing, or simply spending weekend time with children and friends.

The house sits on a large wooden deck supported by piles, an arrangement which, aside from solving the problem of the beach's slope, protects the building from the water. Thus, the deck is the only physical barrier between house and sea. On the western side, which is the least vulnerable to the heavy winds, a swimming pool has been built into the deck. Following the pattern of the exterior, wood and cane were used to finish the inside. All the rooms are on one level, but areas at either end of the house under the high roof have been converted into two small lofts that serve as multi-use spaces.

> The entire complex makes use of natural materials, with wooden framing and walls, a reed roof, and an outer perimeter wall of cane.

Custom-built furniture makes optimal use of the space. The rustic decor includes different kinds of natural fiber mats that, in the absence of partitions, also mark the boundaries of various functional areas. The result is a peaceful, welcoming place that contrasts sharply with the spectacular tides of the mighty ocean.

Wood, specially treated to withstand the area's weather conditions, plays the leading role in the structure.

Natural materials and textures, an austere palette, and simple decorative elements shape the very individual, fresh, restrained style of the interior.

An S-shaped floor plan results in interiors reminiscent of a large ship, very much in harmony with the geographical setting.

Three key elements—natural textures, light, and a neutral decor—combine perfectly to achieve a warm, welcoming ambience.

A sensory exercise

Nature is found in every corner of this house, in which glass walls create the illusion of living outdoors surrounded by the ocean.

Puerto Vallarta, Mexico

Self-contained spaces, natural materials, and light as a source of inspiration–these are the three elements used by the architect to build this unique complex in Mexico on an octagonal wooden deck. In the center, surrounded by water, a strategically placed circle gathers the moon's ephemeral reflection.

The house was designed to satisfy the senses of hearing, sight, and touch, through silence, the beauty of the surroundings, and the use of the area's common natural materials: wood, reed, and cane. All are present in the three subtly-curved structures which together create a truly unique dwelling. The circular deck provides a visual solution, since its layout makes it possible to virtually surround the whole structure with water. At the same time, it affords splendid views across the lagoon and the nearby ocean from any of the windows. To heighten the effect, the walls of each building are made of glass and each building is functionally separate. The largest cabin contains the living room, kitchen, and service area. Another cabin houses a studio with the master bedroom and bath, while the third, with two small bedrooms and baths, is reserved for guests.

> There is a continual physical and visual link between inside and outside, which is reinforced by the materials employed.

The wood used throughout the interior of the three buildings and the octagonal deck unites the various spaces and forms a small central plaza which becomes the heart of the complex.

The absence of superfluous objects, such as paintings, antiques, or other decorative elements results in a simplicity bordering on the austere. Nothing must distract from contemplation of the scenery. The plain furniture is custom-designed and made from wood that is very similar to the wood used for the buildings.

The water in front of the house presents an ever-changing, uninterrupted view of the horizon. In one of the most protected areas of the garden, which reaches all the way to the shore with practically no indications of human presence, a small garden provides exotic vegetables and other produce.

Very typical, natural local materials such as wood, reed, and cane are responsible for the style of this geometrically-centered house, where water is always present.

Achieving this structure's unique shapes was not an easy task. The octagonal designs that distinguish the complex were obtained through a slow, meticulous building process. The result: three cabins with undulating curves and light touches that respect the natural surroundings.

The circular deck, made of the same material as the floors, is used for open-air dining. The roof in the outdoor dining area is made of split cane, which provides protection from the sun but lets the light through.

The furniture was custom-designed for optimal utilization of space, employing wood similar to that of the structure. Its light, simple lines support the leading role played by the architecture.

Between sea and sky

This ranch house, located in a eucalyptus grove with an amazing variety of nesting birds, provides a pleasant place to relax

Punta del Este, Uruguay

An Atlantic Ocean setting in Uruguay, as wild as it is beautiful, is perfect for this structure. The owner is Eduardo Pla a digital artist. Jorge Luis Pesino is the decorator. He designed the attractive interiors, demonstrating that one does not need to spend large amounts of money to have a warm, welcoming environment.

The original site was intended for use as a storage area. Thanks to improvements made during the refurbishing, it has become a typical ranch house, straightforward in appearance and successfully integrated into the natural landscape.

The house, built on a wooden deck raised 40 feet off the ground, has two components. The first, with a wood facade, contains the dining room and kitchen on the ground floor and a bedroom and bath on the upper floor. The second contains a living room, completely enclosed by glass walls, with a ceiling of split guincho, a cane that grows in Uruguayan lagoons and was used by the area's indigenous tribes for construction because of its impermeability.

> The space between the ground and the surface of the wooden deck was concealed by a perimeter wall of untreated roundwood.

Loungers for sun bathing are placed alongside a small pool neatly installed in the deck.

The wood, treated with used motor oil and gasoline, takes on a tone which blends perfectly with the surroundings.

The interior of the house contains only a few light-colored pieces of furniture. This minimizes clutter, an important factor in any residence with limited space. Color is supplied by various decorative elements.

The result is a dwelling on the shores of a vast coast, offering practically deserted beaches and none of the urban amenities, but plenty of peace and rest.

The somber color selected for the walls, the floors, and even the upholstery of some of the furniture is broken by decorative pieces such as these two armoires in a dark wood with vertical lines, located beside the stairs.

The decor features little in the way of furniture, so there can be room to breathe. The living room windows provide a constant flow of light, diffused by white window shades. The colors utilized enhance the brightness of the room.

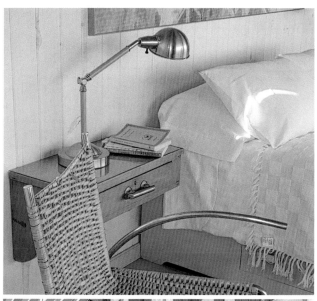

The bedroom is sparsely furnished, but everything there is highly functional. To optimize the use of space, a small unit attached to the wall serves as a nightstand. The room is done in the same austere colors as the rest of the house.

The outside area has been equipped so the occupants can enjoy the open air during the hottest months. Wood in this part of the world is indispensable. Note how the roofing results in an attractive play of light and shade.

The compact size of the house makes any solution to the problem of space welcome. The kitchen is located in a corner of the living room, separated from it by a multi-functional wooden divider which can be used for shelving or as a counter.

Alone beside the sea

Proud of its isolation, this simple structure is a fine refuge for forgetting everything except peace and quiet.

Punta del Este, Uruguay

The geographic conditions–a heavenly spot in Uruguay with more than seven miles of virgin sand–and the climatic conditions–occasional strong winds and periods of unbearable heat–provide the setting for this dwelling.

The house has the ocean as its only companion because an isolated site was deliberately chosen for it. The architect has perfectly integrated the dwelling into the surrounding landscape while also accommodating the climate. The simple, controlled, straight-line architecture permits access to the outside from any place in the house. This results in a fluid, unbroken spatial distribution.

The outer walls are faced with a traditional rustic stucco, which requires no maintenance. A color matching the sand was chosen to integrate the house with the landscape.

The dwelling is built around a large central interior space containing the living room/dining room, porch/terrace, patio, and swimming pool. This area becomes the focus of attention at times of high winds because it is more protected than the other rooms and serves as a refuge. The two towers, one at each end, equidistant from the central area, contain bedrooms and baths.

Lapacho, a local wood as hard and strong as teak, was used for the exterior trim and the terrace furniture, because it can withstand high temperatures and humidity. The interior is decorated with furniture in a range of styles, with the emphasis on the eclectic and personal.

The tower roofs are made of quincho, a common local cane which grows in the lagoons. It keeps the interior warm in winter and cool during the hottest months.

A curtain separates the kitchen from the dining area. Both rooms have been kept quite simple, with minimal furniture. The decorative restraint and the color combinations result in pleasant, welcoming, tasteful settings.

The roof of the side porch, which runs the length of the house with spectacular ocean views, is made of eucalyptus branches, a material used in many local buildings.

The windows in the guest bedroom are portholes trimmed with blinds and curtains. The chrome-plated brass beds are Armenian, the mattresses Indian, and the printed cotton curtains and slipcovers on the sofa, Syrian.

The master bath occupies a corner of the master bedroom. Under the large windows, whose blinds subtly filter the light, and adjacent to the tub, is an iron frame with a mattress and matching cushions that sometimes serves as a sofa.

The southern lighthouse

This house, a mermaid washed ashore, is built between the ocean and the land, alongside an old lighthouse.

La Rochette, France

The architect had two missions for this project: make the house big and make the house functional.

With this wide creative latitude, he drew up the plans for an asymmetrical dwelling surrounded by a distinctive wooden deck which would serve as separate outdoor porches while raising the structure off the beach. The architect's choice of natural materials like stone, wood, cane, and wattles reinforced the utilitarianism of the exterior.

The three-story dwelling, located in France, employs novel decorative and constructive elements. On the lower floor is the guest wing, built in accordance with the natural topography, with direct access to the garden. To make optimum use of the site's natural illumination, several windows, employed as structural elements because of their special high-strength glass, were placed in an upper wooden deck.

The ground floor is a light-filled space on two levels, containing the dining and living areas. The higher part, used as the dining room, contains a large wooden table and four simple chairs painted red. The living room has two built-in bookcases extending the full height of the room on both sides of one of the large windows. In front of the fireplace are two armchairs and a large sofa, all upholstered in white canvas. In the center is a large low table made of tropical wood that breaks up the otherwise stark white layout and highlights the gray of the stone flooring.

> This asymmetrical house is surrounded by a wooden deck that constitutes the flooring for the porches.

A wood and iron stairway leads up to the master bedroom. This room is flooded with light from a large window, which also provides ocean views. Along the side wall, a discrete opening allows visual contact with the living room.

Geometric volumes, constructive solutions full of imagination
and an intelligent take on how to make the best use
of space are this project's best qualities.

The differences in grade, far from being a problem, were taken advantage of by stepping various areas. The result is a building with interesting architectural solutions and visual charm.

The views from all of the porches are extraordinary. The feeling of freedom, with the ocean extending to the horizon, is absolute. The wattle screen covering the porches is topped by lapacho, a local reed, which combines with the wood and stone to produce an irresistible overall effect.

The porches have flooring made of local wood. Properly treated, this material can withstand rain, wind, and the hot summer sun.

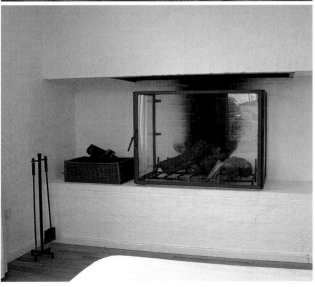

A modern glass-and-iron fireplace is recessed into one of the walls of the master bedroom. The combination of materials creates a piece with clean lines that keep it from being visually overwhelming.

Visual dialogue

Pure geometry and straight, well-defined lines
form an atypical structure in the best tradition
of rationalist architecture.

Santa Barbara, California, USA

The architectural basis for this house, right on the California shore, is the establishment of a constant connection with nature through a simple visual dialogue. Avoiding any excess, the complex unfolds around a cross-shaped swimming pool, placed at ground level to create a chromatic relationship with the ocean, which is framed in the living room's magnificent picture windows.

The play of perspectives developed by the architect in this building takes on an almost spiritual prominence that can be perceived as soon as you enter. The light, wood structure with a canopy of split–cane slats completely merges and perfectly balances the interior and exterior. The floors, on the same level and employing a single dark wood, enhance the continuity between both areas. The flooring contrasts beautifully with the sand color of the walls and the pale cream trim.

In the main room, like a rectangular loft with two spectacular glass walls, is the living space. This simple feature personalizes a building which would otherwise appear to be an elongated cube. It also permits the independent placement of the other rooms, the kitchen/dining room and bedroom with bath, on the sides. An efficient linear distribution with great stylistic power enhances the effectiveness of the decor, which is based on an elegant, neutral white. In the kitchen, this color is diffused and takes on the texture of solid wood, like that used in the floor

> The use of glass walls in the living room and the room's carefully-planned orientation create an incomparable pictorial metaphor.

and in some of the furnishings, including the table in the center of the living room and the bathroom vanity.

The only notes of color are provided by a handcrafted rug in earth tones as well as all the chromatic splendor of the Pacific.

The physical boundaries between inside and outside have completely disappeared. The passage from one to the other is immediate, and the visual play that is created in every corner of the house works a kind of magic.

There is a feeling of continuity inside the house because the same atmosphere permeates the areas used as kitchen, dining room, and living room. It is an elongated space, bound by two parallel glass walls that allow the constant entry of light from the outside. The white walls and white furniture augment the luminosity.

The bath was placed independently on one side of the house, and was furnished with only the barest essentials. A compact vanity with sober, straight lines incorporates a mirror and basin, thus conserving space.

To make the best use of the space, the kitchen solves the problem of storage with a solid wooden shelf that leaves dishes and other tableware and utensils in plain view.

Given the simplicity of the furniture, great care was taken with the choice of decorative elements. The exquisitely designed tableware, embossed with a fragment of a poem by Pablo Neruda, blends with the rustic-style place mats and simple blue glass tumblers.

Bathed in light

Intense white prevails in this seaside structure, whose watchword is simplicity.

Wellington, New Zealand

Light textures; wide, interconnecting spaces united by a single tone, white; and simple, spare decoration are the keys to this house located in a setting of spectacular beauty in New Zealand.

The woman responsible for giving expression to the decorative dreams of the owners and for assuring that these magnificent interiors receive their full measure of natural light is decorator Laura Orcoyen.

Two porches dominate the main facade of this enormous white, wooden house. The rooms were carefully distributed between the two floors: the lower floor is used for living space and the upper floor houses the bedrooms and bathrooms. To provide a feeling of continuity between the interior and the exterior, umbrellas scattered around the terraces were covered in the same fabric used on the living room sofas.

The salvaged brick used for all of the walls was painted exclusively in white. This provides a powerful sense of roominess and brightness, the light spreading through discreet, comfortable interiors and leaving little or no room for superfluous adornment.

This minimalist approach accounts for the fact that only essential furniture is found inside. Two large sofas, facing each other in front of the fireplace, and a few tables of white-stained wood are the only pieces in the living room. The dining room, which is adjacent to the living room, is also simply furnished, with an oversized table and white wicker chairs along the wall, beneath a series of black-and-white photographs. The rest of the rooms follow the same principle of moderation.

The chromatic continuity of the facade is broken by the blue-green tone of the large gateway, inspired by the color of Tiffany jewelry boxes. This shade is also found on the sofas' throw pillows.

As a whole, the house is as an attractive mixture of austerity, harmony, and tranquility. A welcoming place from which to enjoy gazing at the ocean.

Clean and simple lines,
rooms purged of anything superfluous,
and an attractive austerity define the
interiors of this spacious house

The narrow confines of the bath do not detract from its comforts. The jacuzzi placed between the two windows is the most striking element in a room in which white is the principal color, as it is throughout the dwelling.

Photographer Vicky Aguirre created the images of nature used in different parts of the house. This evocative photograph, framed in white, hangs over the fireplace like a mirror, because its transparency reflects the back part of the room.

Simple glass vases, one of the few decorative elements in the room, adorn the long dining room table. In the background is part of the white wooden staircase leading upstairs.

The wood plank flooring on one of the porches is painted white, like most of the furniture inside. The striped cloth used to cover the cushions and the rope securing the curtain add to the maritime feel of the house.

Mediterranean inspiration

The Atlantic is at your feet, but the spirit is markedly Mediterranean, an example of how styles can be extrapolated without losing the enchantment.

Punta Piedra, Uruguay

The owners of this house, in love with the Ibizan style, were inspired by their memories to bring the typical architecture and character of the Spanish island to life.

This building bears little resemblance to the small, drab house they found on the beach at Punta Piedra, Uruguay, a decade ago, when the area was practically uninhabited. The meticulous renovation of Can Torres (a name, reminiscent of the Balearic Islands, acquired during the refurbishment) resulted in a structure of elegant beauty, with a turret and a terrace from which the Atlantic's color changes can be enjoyed—all in homage to the Island of Ibiza, where the owners had spent so many summers.

The need for additional space made expansion of the dwelling unavoidable, although the work itself is practically undetectable: the original building and the new construction are perfectly complementary, as if they had been created together.

Later, again for reasons of space, architect and designer Mario Connio drew up the plans for a definitive expansion, and so we have the building as it is today. Added onto one corner was a new section joined by a stairway with charmingly simple lines. The addition included an attractive interior patio set off by columns copied from an Ibizan church. The swimming pool is located there, adjacent to the living room dining room, kitchen, and a bedroom suite.

The simplicity of line is also maintained in the interior decor and furniture: old, inherited pieces with great sentimental value, Mediterranean-style furnishings, and mementos accumulated over the years. The intense light is amplified by the white walls throughout the house. And as a final decorating note, the interplay of light from the carefully thought-out distribution of windows brings the exterior panorama inside.

Sober, restrained decor enhances the calm, peaceful atmosphere of the interior. The rooms are dominated by a neutral white, broken only by carefully-chosen touches of color and the play of light.

The original building is joined to the addition by a stairway with classical lines, a stairway which is enormously attractive in the context of the large space. The stairs provide a connection between the old and new sections, as well as effective protection from the sometimes very high wind and the indiscreet glances of curious passersby.

In a corner of the master bedroom is a work area with English furniture. The solid wood of these decorative pieces contrasts with the white walls filled with photos and mementos.

An old fishing buoy provides a unique finishing touch to the handrail of the staircase leading to the second floor—another reminder that the Atlantic is right outside.

A return to tradition

Made entirely of eucalyptus wood, this cool, fancy-free building appeals to the most basic senses.

Los Cabos, Baja California, Mexico

Generous dimensions and an extended horizontal line are two of the most obvious features of this house in Mexico.

A long, white wooden porch brings out the soft turquoise of the facade and leads to a house with very personal, colorful interiors.

One of the key decorative elements is the large, loft-like central space, in which the various areas flow together naturally and continuously. The dining area adjoins the living room, where a quiet work nook has been set up. This, in turn, is next to a canopied four-poster bed, which is opposite an old claw-footed bathtub.

The cold stone, exposed in several places in the living space, is softened by the warm tones of various fabrics and by decorative elements inspired by the pre-Columbian era.

Except for the wooden cubes used as stands, which were created by the owner-designer, all the wooden furniture reproduces traditional local designs and was made by local artisans.

The armchairs are upholstered in colorful, Santa Fe-style fabrics, while the floor is dotted with Mexican throw rugs featuring Aztec motifs, setting off the bright lemon-colored sofa.

In a room adjacent to this space is the master bedroom. Blue dominates here, contrasting with the natural wood of the rough-hewn, unvarnished bedposts and the Mexican rug. Thick, elegant French curtains frame the windows. The color combination helps break up the intense coldness of the blue and gives the space added warmth.

Style, color, and fantasy combine in a dwelling where subtle, elegant interiors shape an imaginative contemporary space without relinquishing the use of the country's traditional materials and forms.

Chosen for its brightness, warmth, and functionality, wood predominates inside and out.

The generously-proportioned porch surrounds the entire building. It can be enjoyed during the hottest months. Sparse, well-chosen furniture sets the stage for this basic, utilitarian outdoor spot.

Wooden cubes, used as stands, are scattered around, enhancing this unique space. Simple furniture was chosen, leaving the fabrics and various pieces to add a note of color.

The bath is painted entirely in sea green. With a simple foundation and a pinch of imagination, a highly personal atmosphere was created. The success of this room lies in the conscious mixing of different styles. The vanity and the rest of the items were acquired at various antique auctions.

Domesticated shapes

Arrogant, audacious, and flamboyant—this house in no way tries to conceal its beauty from anyone who wants to look.

San Juan, Puerto Rico

Located beside the ocean, on the southern coast of Puerto Rico, this house takes its inspiration from the shape of a large bird cage. The structure, which recently won an architectural prize for design, is a spectacular, immensely appealing building.

Large proportions, exquisite content, and a meticulously-detailed architectural plan result in a complete package that invites easy living. No conventions, no worries. This is a lesson in balance and beauty, in which a pact was signed with nature, allowing the inhabitants to enjoy everything around them.

Visually, the wooden structure goes beyond the attractive and unique. The house is built of superimposed cubes that form the different floors, based on an open, vertical concept in which the more enclosed modules become rooms for the inhabitants while the more open ones, painted blue, nearly always The modules are interconnected, so the interior and the exterior converge at all times, while light and air pervade the various rooms unimpeded.

The unique lines of the building, inspired by traditional bird cages, leave no one indifferent and demonstrate that beauty and practicality can coexist.

Local wood, carefully treated and sometimes stained, is the predominant material. In addition to the frame, it has been used in the flooring, roofs, certain walls, and much of the furniture.

A large, unpartitioned, undelineated living space occupies the first floor: living room, dining room, and kitchen are one, with a terrace that is largely covered, and a small balcony. The bedrooms and bath are upstairs.

The utmost care was taken to achieve an immutably modern residence where the living is easy, thanks to agreeable order and strict disorder. Pleasure rules!

Behind the innovative design are comfortable interiors which may be sparsely furnished and without luxury, but which are conducive to rest and relaxation.

The kitchen, which shares space with the living room and dining room, is at one end of the large multi-use area. Simple to a fault, its materials and color combinations consolidate the many facets of the space.

A solid, wooden staircase leads to the upper floor, which houses the bedrooms and the bath. The simple, well-defined lines give power to the architecture without detracting from the decorative background.

The interiors were laid out simply. The two-story space makes the decorative elements more important, with the visual integration of the sloping roof and the special chromatic treatment afforded the doors and windows.

Following the same guidelines as the rest of the house, the bath maintains simplicity of form. An antique bathtub, set between two sturdy, rough-hewn pieces of timber, balances the placement of the fixtures, which are the same color as the tiles--a clever solution that creates the illusion of space.

Part of the landscape

Refined architecture and restrained design in a
welcoming dwelling with a mixture of styles.

Carmel, California, USA

A weather-beaten facade whose texture provides a pleasing aura of vertical dynamism conceals a surprising interior that combines comfort and imagination. The exterior, atypical for the beach, may give a false impression: the austerity contradicts the warmth and comfort inside. The technical and decorative solutions are exquisite, yet highly practical.

Many large windows provide brilliant natural lighting–the absence of curtains and blinds making the landscape seem like a canvas on the wall–and an open layout is rewarded by the play of perspectives between the levels and through the glass interior walls. The various rooms in this California house are formed by a succession of cubes, all in constant touch with the exterior. At the same time, in spite of its visual contact with the surroundings, the building is structurally self-contained and surrounded by a wood plank fence: all in all, a disciplined, compact architectural design.

The straight lines and soberness of the exterior, a part of the landscape, totally belie the imaginative, charming interior of the house.

The main construction material is wood. Varnished or unfinished, it is largely responsible for the warm ambience of the house and is the perfect complement to other materials, like the flagstones covering the downstairs floor or the stainless steel in the kitchen fittings.

Upstairs, a walkway with large windows on one side and a three-dimensional railing on the other leads to the bedrooms. A glass shower divides this into two areas. Each section contains a bed placed flush against the window with a stylish canopy that doubles as a translucent window curtain.

Interesting combinations of styles result in eclectic, personal spaces. The exultant modernity of some rooms contrasts with areas decorated in a more classical, rustic, and restrained style.

Nature is frequently a part of the decor. In the bathroom, for example, the customary mirror is replaced by a window, which looks like an original, improvised landscape painting.

An unusual combination of materials is a fundamental part of this house's decor. The canopy for the beds is made of copper, steel, wood, and a gauze-like fabric.

Inspired by the coolness of high-tech, this unusual three-dimensional railing totally shuns the functional in favor of a striking, innovative, nearly invisible design.

Reuse is often the solution, as in the case of this mirror framed by slender steel tubes and burlap, which provides character to a house where imagination and a little wildness reign.

A terrace overlooking the sea

This exquisitely simple old building with whitewashed walls guarantees an uncomplicated, quiet, natural way of life.

Essaouira, Morocco

This house in the medina of Essaouira, Morocco, has three multi-level stories. The renovation respected the original structure, retaining the frame while transforming the venerable building, battered by the passage of time, into a warm, welcoming place with all the enchantment, power, and splendor of the past.

As in most of the quarter's traditional buildings, the thick, whitewashed stone walls, contrasting with the intense blue of the Atlantic, are simple, pure, and unassuming. This architectural restraint is maintained in the cool interiors, where opulence is conspicuous through its absence.

Almost every room in the house opens to the outside, where freedom is tangible and luminosity, fueled by the whiteness of the exterior, is pervasive. Inside, the rooms are also painted entirely in white, set off by black-and-white floors in a chessboard pattern, surrendering unconditionally to the light and the austerity. The efficient layout and the construction methods and materials make the most of the available space. The multi-use furnishings, nomadic and diverse, are spartan and plain. A sober ambience, which in the bedrooms is almost monastic, is the guiding principle of the decor.

With a tower of the neighboring house in the background, the whitewashed terrace offers a magnificent setting for sunbathing on the attractive white cushions.

This is a modest structure that seduces with its exceptional restraint and purity, a place of rest, an unostentatious building that avoids the superficial, a dwelling in which the inhabitants' lives are conditioned by the nearness of the ocean.

The owners' extroversion can be seen in the mix of styles: a small Arabic tea table, a functional chair, a straw magazine rack, a wooden chest, a tiled fireplace, the chessboard metaphor.

The kitchen, which contains only the bare essentials, employs the same color scheme as the living room.

The bathroom, recessed into the wall, occupies a minimum of space. The outer surface of the white-tiled shower has a mosaic of small, triangular, decorative ceramic pieces in the same design used on the living room fireplace.

A simple pulley system and a basket made from tire rubber, purchased at the local marketplace, are used to move things around without climbing the stairs.

The neutral white is broken by a chessboard pattern of Moroccan ceramic tile that covers the floors and decorates the baseboards and the bases of columns.

To lighten the rooms, some of the heavy doors, such as those in the bedrooms, were replaced by light, ethereal, white cotton curtains.

A token of simplicity

Camouflaged in an arid landscape, this house beside
the Aegean Sea reflects an austere, meditative lifestyle.

Cyclades, Greece

In a harsh, elemental natural setting, built for a way of life based on simplicity, this house in the Cyclades (Greece) is very typical of the local architecture.

It is small and its construction took advantage of the features of the terrain. The use of stone and the thickness of the walls assures coolness in a location where extremely high summer temperatures are common. This is also why, for example, doors and windows were kept small and wooden shutters were installed. The interior walls were whitewashed to afford protection from the heat and to make the house brighter as the light is reflected and diffused.

One first notices the two small wood–framed, cane–roofed porches, which make efficient use of the slope. They have only the bare essentials for outdoor living and enjoy incomparable views. Inside, the small rooms, decorated in a rustic Mediterranean style, are efficiently laid out to make the best use of the space. The furniture is simple and basic, often secondhand. In addition, features which reflect the house's rustic atmosphere are employed, such as the small niches in the walls which hold books and the stone shelves with wooden brackets.

A typical Mediterranean landscape surrounds the house, which is practically invisible on the shores of a sea nearly as old as the lifestyle it suggests.

This house reflects an attempt to return to a natural lifestyle, in which comfort means basic comfort, without the luxuries of the faster-paced urban environment.

On the porch, a stone bench, which is built into the facade, provides a resting place. When the sun sets, it is illuminated by a method as novel as it is rudimentary.

The combination of whitewashed walls and irregular,
unpolished stone floors makes for a characteristically
Mediterranean interior.

The broom without a
handle and the wood
and esparto chair are
typical items.

The house has practically no electrical appliances. For cooking, the residents use a small gas stove, which takes up little space and is easily moved.

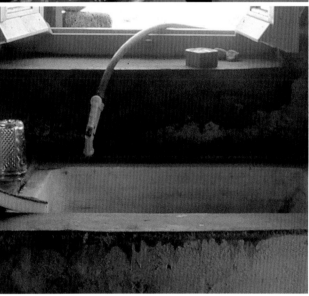

An old marble sink under the window is used for washing dishes. The water comes from a hose which is also used for an after-beach outside shower.

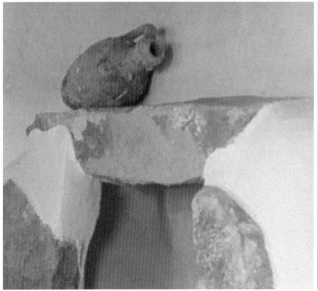

As in most houses in the Greek islands, the fireplace is of simple mortar. This one is built into the whitewashed stone wall in a corner of the living room. A small amphora vase purchased from a local antique dealer is the only decorative piece.

A passion for what's natural

This seaside refuge has Mediterranean views in every direction and enjoys an unparalleled setting that is an open invitation to relax.

Panteleria, Italy

On the Italian island of Panteleria, off the coast of Sicily, this formidable, emphatic building seems to be sculpted from the rock. Its only companions are the intense blue sea all around, the green vegetation, and the sky. The senses, especially sight and smell, come to the fore. A dazzling, lively explosion of color fills every room; the smell of sage, lemon, and spike lavender flood every corner of this dwelling, open to the sea and nature.

Its architecture is typical of the region: pure, simple lines and solid stone construction which is often virtually camouflaged, since the structures are planned and built with the terrain in mind.

The thick stone walls, natural textures, and intense colors produce quiet, comfortable spaces designed for moments of leisure; there is really no place for luxuries or superfluous items. A certain Oriental flavor dominates in the materials, fabrics, accessories, ambience, and the simple layout of the room. The austerity of the facade contrasts with the cheerful colors inside, but even here there is no extravagance, to demonstrate that the things that are important can be very simple and that it is possible to live with only what is absolutely necessary.

On the front porch, where a pleasant area for relaxation has been created, the hard rock and the leaf roof contrast with the lightness of the hammock and the white curtain that serves as a divider. An unusual natural fiber carpet becomes an improvised table for a refreshing glass of lemonade.

Rooms that connect naturally and freely, sparse furnishings, few decorative objects–what there is comes from the owners' travels–enhance this refuge's personal, down-to-earth atmosphere. The result is a home full of charm and the sweet air of antiquity. It is as if time had stood still and intense enjoyment of life were the only objective.

Fresh, vital colors, natural materials and textures, and a purposeful austerity combine to create a home that champions a simple lifestyle, in which time passes to the rhythm of the sun and the sound of the sea.

Facing the sea, at one end of the house, is a porch suitable for get-togethers and conversations with friends at any time of the day. The cane roof provides ample shade even on the hottest days. The only furnishings are a large stone banquette covered with cushions, a colonial-style bench, and a hammock.

Using a traditional local technique, the inside of this laundry sink was covered with decorative ceramics. The old enamel basin beside the sink was picked up at an antique shop.

Most of the rooms in the house use fabric in place of doors. This blue cotton curtain separates one of the rooms from the outside, an effective means of providing shade from the strong sunlight.

Naturalness and simplicity are present in all the details. Healthy food, earth tones, and a Mediterranean touch can be found in every corner.

Architectural synthesis

Perched among the rocks, enjoying a
breathtaking panorama, this house is an homage to nature.

Ibiza, Spain

This house, atop a promontory on the island of Ibiza, Spain, is the product of a dialogue between nature and architecture, resolved in light, bright spaces physically integrated into the rock on which it rests.

The natural setting dictates the placement of the large windows, which are strategically located to offer fabulous views. Another feature connecting the terrain with the everyday life of the household is the arrangement of the main terrace, which has no railings or anything else which might obstruct the view of the landscape. Its only decorative element is a large, immaculately white sofa.

> Ensconced in a breathtaking natural setting, this dwelling of clean, geometric lines, sometimes disappears into the landscape and shows how architecture and nature can coexist without damage to either one.

The predominance of the large windows, framed in black aluminum which is visually linked to the stripes in the flooring, is broken only by the bright white hue chosen for its decorative purity and its promise of lively and attractive contrasts. All this is combined with the strong natural light, resulting in an explosion of elegance and simplicity.

The placement of the furniture and decorative objects is deliberate. The interiors follow the architectural guidelines that define the house's exterior. Nothing is superfluous or excessive.

Harmony is maintained throughout. In the dining room, a set of large earthenware jars, opposite a glass table set on four white concrete columns, is aligned next to a window framing a stony precipice.

In this synthesis of nature and architecture, each room relates differently to its surroundings and creates, according to the changing light and season, different spaces which dissolve the boundaries between interior and exterior.

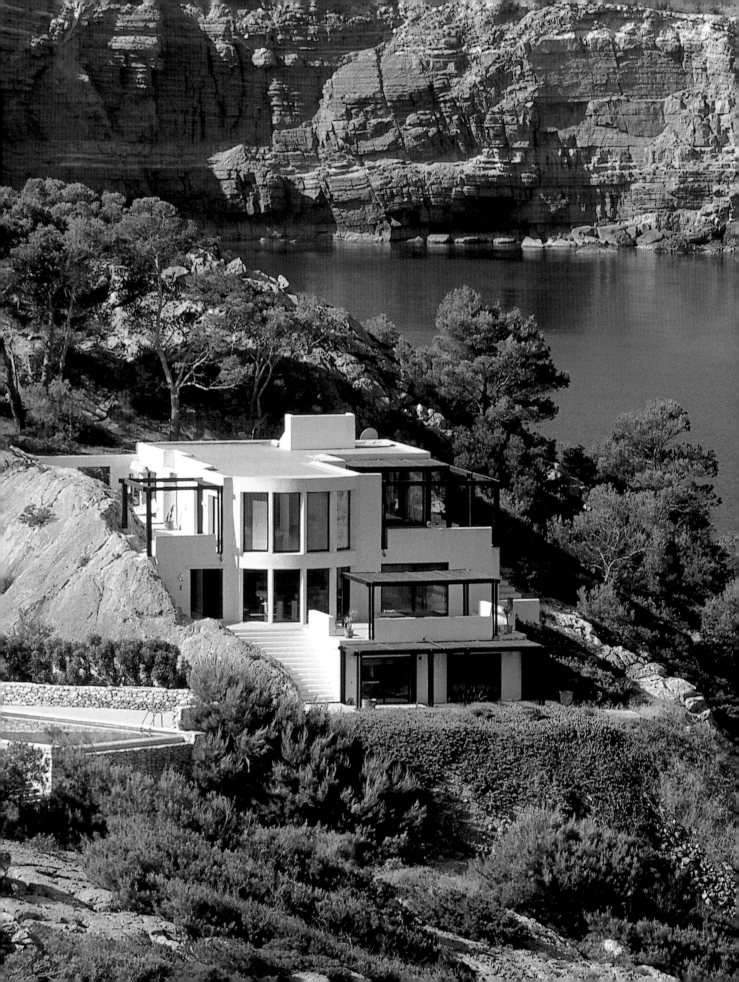

The interplay of landscape and architecture is spectacular.
The refined, sun-drenched rooms with their clear, sleek lines are the hallmark
of this structure.

Taking advantage of the slope, several terraces cling to the house. An aluminum framework supports a split cane roof that provides protection from the strong Mediterranean sun and visually defines the space, which is open to the sea and which contains only essential items.

The master bedroom, at the end of a light-filled passageway, retains the sensation of diaphanous spaciousness found throughout the house, thanks to the continuity of the white walls. This shade is also used in the bath, which is separated from the bedroom by a wood-topped divider.

The architectural solutions make full use of the stone on which the house is built. The guest bedroom has a bathtub shaped like a natural pool.

White geometry

This house, with its conciliatory spirit, molds the light to its own will.

Ibiza, Spain

This structure with a markedly Mediterranean air, whose form and content follow the dictates of Ibiza traditional architecture, is hidden in thick, lush vegetation. One of the goals of this project was to return to nature the space that is so often wrested from it.

The simplicity of both the exterior and the interior decor are dominant features in this house of thick, whitewashed walls that reflect the brightness of these lands. Light is supreme thanks to the numerous openings cleverly placed in each wall and the white tone that predominates both inside and out. In many parts of the house, this purity is a controlling factor and the ideal backdrop for the array of decorative objects which humanize the space.

White and radiant, the building, which is typical of the island's houses, stands proudly amid the chromatic vitality of the surrounding trees and plants.

The building has a powerful, well-defined, geometric shape with a white facade that brings out the rich, contrasting color of the surrounding trees and bushes.

Inside, where the white tone is maintained to provide continuity, the rooms are laid out on two floors open to the outdoors and perfectly integrated with the surroundings.

An open, welcoming room contains both a large living space and a dining area. In front of a fireplace, set between two large windows that visually connect the interior and exterior, is a distinctively Arabic spot for relaxation, conversation, or coffee. The kitchen is next to this large area, up a few steps and through two door-like openings in the wall. This design solution creates an attractive visual interplay, in which light is a key factor.

In a corner of the living room is a flight of stairs that leads up to the second floor, an utterly austere area dominated by white, containing the bedrooms and baths. The master bedroom has three clearly differentiated, but perfectly integrated, areas on different levels, each separated by a couple of steps.

Sobriety and discretion are the outstanding qualities of the interior decor. There are no dramatic or pretentious elements; each object has a specific reason for being there. There is no room for frivolity, only for tranquility and harmony.

Collector's items, objects from the island's craft shops, and other decorative elements are scattered throughout the rooms, where the brightness of the white walls, enhanced by sunlight, envelops everything.

Clean lines and carefully chosen details
show absolute respect for the surroundings.

The stairway is also whitewashed to create visual continuity. The tone highlights the stone steps and underscores the austerity that predominates throughout the house.

The spirit of the traditional rustic kitchen is reflected in the materials used and in the simple style. Effective construction techniques and elegant decorative touches make it highly functional.

The bedroom, a unique, multi-purpose space, has three clearly delineated, but integrated, sections on different levels.

Locations, Architects and Designers

The beauty of the imagination
Punta del Este, Uruguay
Architect: Martín Gómez
gomezarq@adinet.com.uy
Production: Víctor Carro

Facing the ocean
Punta del Este, Uruguay
Architect: Pachi Firpo
Production: Dos Ríos

With the ocean at one's feet
La Jolla, California, USA
Production: Dos Ríos

Maritime exoticism
Punta del Este, Uruguay
Architect: Mario Connío
m.connio@stl.logiccontrol.es

Natural fusion
Zihuatanejo, Mexico
Architect: Mario Connío
m.connio@stl.logiccontrol.es
Production: Víctor Carro

A sensory exercise
Puerto Vallarta, Mexico
Architect: Mario Connío
Production: Víctor Carro

Between sea and sky
Punta del Este, Uruguay
Decorator: Jorge Luis Pessino
Digital artist: Eduardo Pla
Production: Víctor Carro

Alone beside the sea
Punta del Este, Uruguay
Architect: Mario Connío
m.connio@stl.logiccontrol.es
Production: Víctor Carro

The southern lighthouse
La Rochette, France
Architect: Diego Montero
montero@adinet.com.uy

Visual dialogue
Santa Barbara, California, USA
Architect: Pachi Firpo
Production: Víctor Carro